Introduction

Trees of San Marino was inspired and supported by the Huntington Botanical Gardens Overseers Committee. It is a tribute to the city of San Marino, celebrating the importance and beauty of the trees that grow here. We hope it serves to encourage members of this community to become more aware of the nature and diversity of plants in our built environment. Perhaps it will create a desire on the part of people to plant more trees, the dominant feature of the green landscape.

This book describes over fifty of the several hundred species of trees that grow within the city limits. Choosing which trees to include was difficult, but our general principles were to select those that are planted in public parkways, those of great prominence in the landscape, some of the more curious species, and a couple of trees that merit more frequent planting.

The book is organized so that each tree has an illustration and a narrative description that will help with identification and provide useful information. At the end of each description is a general location in San Marino where an example of this kind of tree can be found. The locations, when possible, are in public parkways and rights-of-way, rather than private properties. The sites were selected so that fine examples of trees from all areas of town are represented. However, no attempt was made to document the entire inventory of street and park trees in San Marino. Following the Table of Contents is a map of the city which identifies these locations.

Many people participated in the publication of this book and their contributions are greatly appreciated. Wendy Stubley located the trees, wrote all of the narrative descriptions, and coordinated the process of illustrating each specimen. Her fortuitous arrival in Southern California while on sabbatical from the University of Guelph, Ontario, Canada, marked the true beginning of this project. By the time she returned home, the text, illustrations, map, indices and appendices were complete. For her work Ms. Stubley was guided by the ideas and encouragement of the members of the Gardens Overseers Committee: Mrs. Richard Call (Chairman), Alice O'Neill Avery, Mrs. Guilford Babcock, Mrs. Frederick Larkin, Mr. Malcolm McDuffie, Mr. Edwards Metcalf, Mr. Dennis Stanfill and Mr. Robert Wycoff.

The fine illustrations of each type of tree are the result of the patient, responsive and artistic labors of Pat Brame, Jacquelyn S. Giuffré and Marianne Wallace.

Countless hours of conversation, site visitation, reading and editing went into the making of this book. In this regard, we have been very fortunate to benefit from the efforts and talents of the following people: Catherine Babcock, Fred Brandt, James L. Degen, Debra Folsom, Ronald Harris, Donald R. Hodel, Myron Kimnach, Clair Martin III, Kathy Musial, Virgil Nichols, Danielle Rudeen, Dave Sivertsen, A. Betty Southam, Guilland Sutherland and John Trager.

We are also fortunate that Ava Moeller, the Huntington's graphic designer, devoted the time to

TREES OF SAN MARINO

by
Wendy Stubley

THE HUNTINGTON

LIBRARY, ART COLLECTIONS AND BOTANICAL GARDENS

Library of Congress Cataloging-in-Publication Data

Stubley, Wendy, 1960-
Trees of San Marino.

Includes bibliographical references.
1. Ornamental trees—California—San Marino.
2. Trees—California—San Marino. I. Title.
SB435.52.C22S257 1989 635.9'77'0979493 89-24724

ISBN 0-87328-129-2

Illustrations by Pat Brame, Jacquelyn S. Giuffré and Marianne Wallace.
Design by Ava E. Moeller.
Printed in the USA by PSP Graphics.

work thoroughly with the Botanical Gardens on the project, developing the layout and design for this book.

Finally, we would like to thank the generations of plant enthusiasts who planted the city landscape with trees and shrubbery. This book is dedicated to those past and present tree lovers of San Marino.

James P. Folsom
Curator of the Gardens

Table of Contents

CONIFER TREES

The Ginkgo Family

Ginkgo biloba . . . . Ginkgo 1

The Pine Family

Cedrus deodara . . . . Deodar cedar 2

Pinus canariensis . . . . . . Canary Island pine 3

Pinus halepensis. . . . . . . Aleppo pine 4

Pinus pinea . . . . Italian stone pine 5

Pinus thunbergiana . . . . Japanese black pine 6

The Taxodium Family

Sequoia sempervirens. . . . Coastal redwood 7

Taxodium mucronatum . . Montezuma cypress 8

The Cypress Family

Calocedrus decurrens. . . . Incense cedar 9

Cupressus sempervirens. . . Italian cypress 10

The Podocarp Family

Podocarpus gracilior . . . . Fern pine 11

BROADLEAF TREES

The Magnolia Family

Magnolia grandiflora. . . . Southern magnolia 12

Magnolia ×soulangeana . . Saucer magnolia 13

The Laurel Family

Cinnamomum camphora . Camphor tree 14

Persea americana. . . . . . Avocado 15

The Sycamore Family

Platanus racemosa . . . . . Western sycamore 16

The Witch Hazel Family

Liquidambar styraciflua . . Sweet gum 17

The Elm Family

Ulmus parvifolia. . . . . . . Chinese elm 18

The Mulberry Family

Ficus microcarpa . . . . . . Indian laurel fig 19

The Beech Family

Quercus agrifolia . . . . . . Coast live oak 20
Quercus engelmannii. . . . Engelmann oak 21
Quercus ilex . . . . Holly oak 22
Quercus lobata . . . . Valley oak 23
Quercus suber. . . . . Cork oak 24

The Sterculia Family

Brachychiton discolor. . . . Queensland lacebark 25
Brachychiton populneus . . Bottle tree 26

The Kapok Family

Chorisia speciosa . . . . . . Silk floss tree 27

The Heath Family

Arbutus unedo . . . . Strawberry tree 28

The Rose Family

Prunus caroliniana. . . . . Carolina laurel cherry 29
Pyrus kawakamii . . . . . . Evergreen pear 30

The Pea Family

Cassia spp.. . . . . Cassia 31
Ceratonia siliqua . . . . . . Carob tree 32

The Loosestrife Family

Lagerstroemia indica . . . . Crape myrtle 33

The Myrtle Family

Callistemon spp.. . . . . Bottlebrush 34
Eucalyptus citriodora . . . . Lemon-scented gum 35
Melaleuca quinquenervia . Cajeput tree 36

The Protea Family

Grevillea robusta . . . . . . Silk oak 37

The Soapberry Family

Cupaniopsis anacardioides Carrotwood 38
Koelreuteria spp. . . . . Golden rain tree 39

The Cashew Family

Schinus molle . . . . California pepper tree 40

The Citrus Family

Citrus sinensis. . . . . Sweet orange 41

The Olive Family

Fraxinus uhdei . . . . Shamel ash 42
Olea europaea. . . . . Olive 43

The Bignonia Family

Jacaranda mimosifolia . . . Jacaranda 44
Tabebuia spp . . . . Trumpet tree 45

PALM TREES

The Palm Family

Brahea edulis . . . . Guadalupe palm 46
Butia capitata. . . . . Jelly palm 47
Phoenix canariensis . . . . Canary Island date palm 48
Syagrus romanzoffiana . . Queen palm 49
Washingtonia filifera . . . . California fan palm 50
Washingtonia robusta . . . Mexican fan palm 51

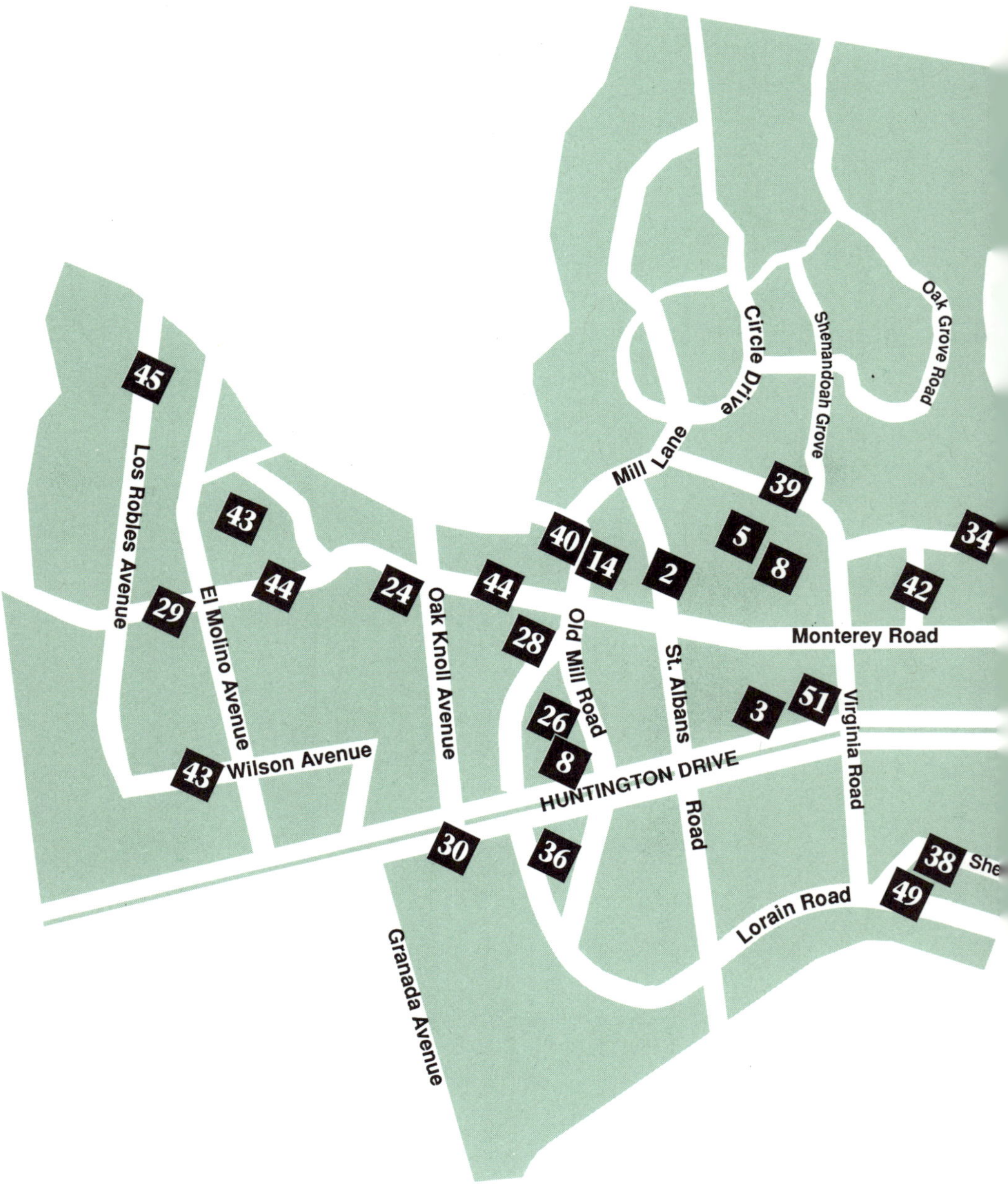

45
Los Robles Avenue
43
El Molino Avenue
44
29
24
Oak Knoll Avenue
44
40
14
2
Mill Lane
Circle Drive
Shenandoah Grove
Oak Grove Road
39
5
8
34
42
28
Old Mill Road
St. Albans Road
Monterey Road
26
8
3
51
Virginia Road
43
Wilson Avenue
HUNTINGTON DRIVE
30
36
38
She
49
Lorain Road
Granada Avenue

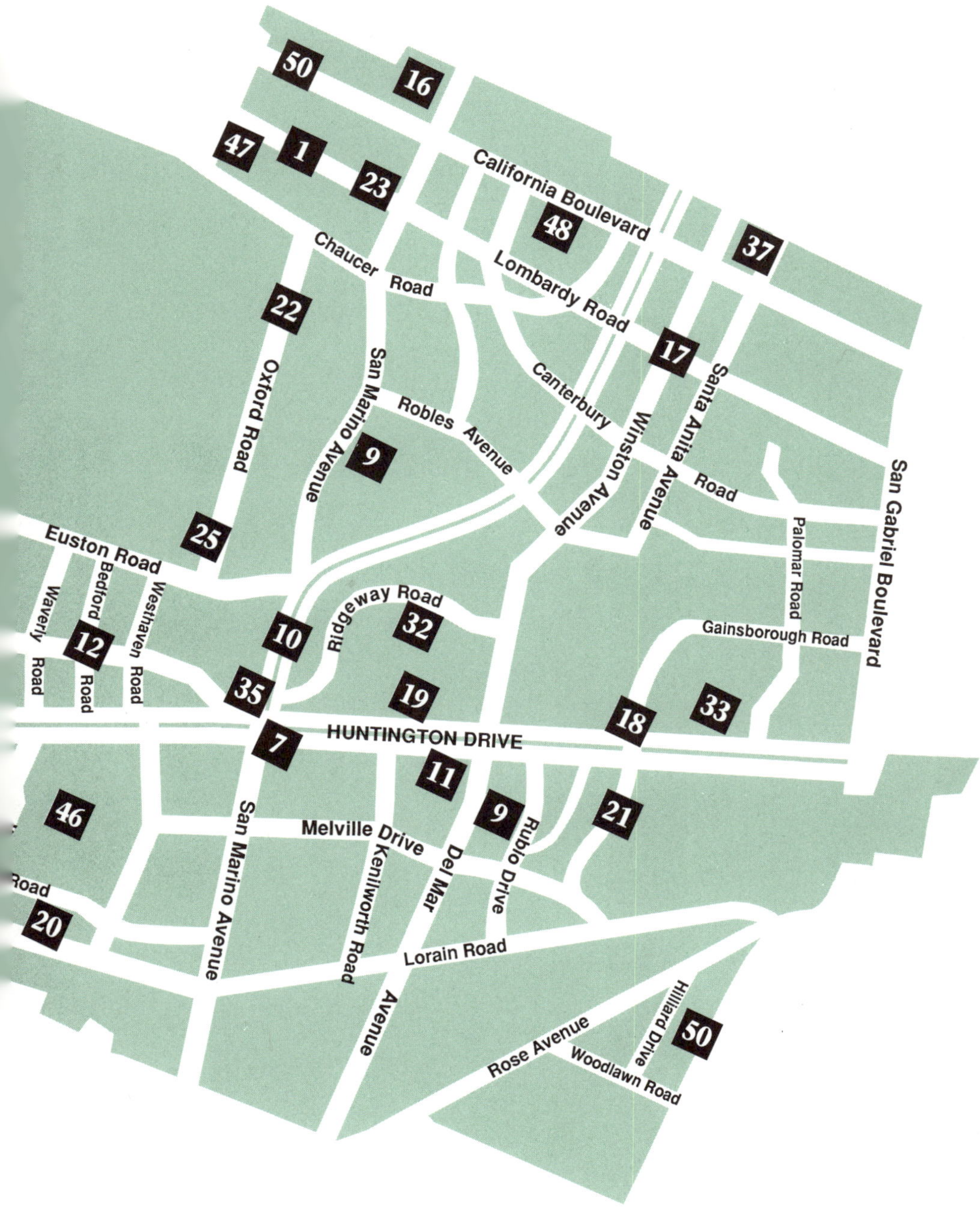
50
16
47
1
23
California Boulevard
48
37
Chaucer Road
Lombardy Road
22
17
Oxford Road
San Marino Avenue
Robles Avenue
Canterbury Road
Winston Avenue
Santa Anita Avenue
San Gabriel Boulevard
9
25
Euston Road
Bedford Road
Waverly Road
Westhaven Road
Palomar Road
Ridgeway Road
32
10
12
Gainsborough Road
35
19
18
33
HUNTINGTON DRIVE
7
11
46
9
21
San Marino Avenue
Melville Drive
Kenilworth Road
Del Mar Avenue
Rubio Drive
Road
20
Lorain Road
Hilliard Drive
50
Rose Avenue
Woodlawn Road

The Ginkgo Family

Ginkgo

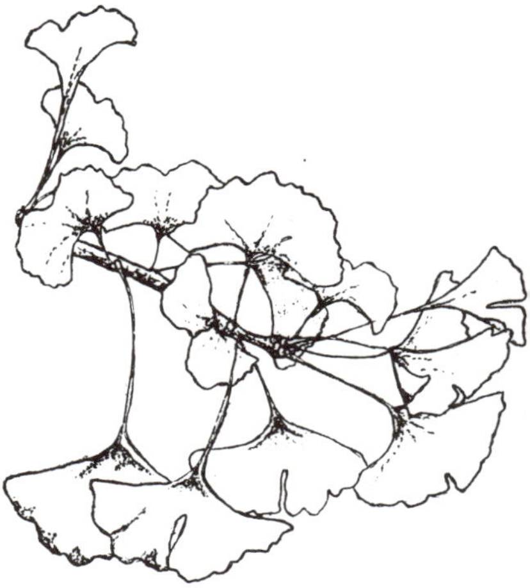

IN SAN MARINO: *Along north side of Lombardy Road between Allen Avenue and San Marino Avenue.*

LOOK FOR: *Distinctive leaf shape (see illustration). Bare branches in winter.*

Ginkgo biloba

This hardy, slow-growing tree is often called a "living fossil" because members of its family are known from fossil remains. The ginkgo is now thought to be extinct in the wild state, but fortunately it was preserved in Chinese monastery gardens where it has been cultivated for many centuries. The ginkgo is more closely related to the conifers than to broad-leaved trees, though this is not obvious from its foliage. The unusual fan-shaped leaves resemble oversized leaflets of the maidenhair fern, from which comes one of the common names, maidenhair tree. The second part of the Latin name, *biloba*, refers to the two lobes which occur on most of the ginkgo leaves. These leaves are light green, but look spectacular in the fall when they turn a brilliant golden yellow.

The ginkgo is dioecious, meaning that the trees are either all male or all female. Since the seed coat from the female fruit gives off an obnoxious odor when crushed, female trees are rarely planted. In Asia, the seed inside these pungent coatings is roasted as a delicacy. The name ginkgo comes from the Japanese version of the Chinese *yin-kuo*, meaning silver fruit.

While young ginkgo trees, especially unpruned seedlings, may be gawky, mature specimens become elegant and well shaped. Their long, graceful branches acquire an attractive lacy look.

Smog, salt air, poor soil, and wind are tolerated by the ginkgo, so it is a suitable tree for Southern California conditions. It grows to only about 50 feet on the West Coast, but can reach heights of 120 feet in more ideal climates.

Deodar Cedar

Of the true cedars grown in Southern California, deodars are the most common. An impressive planting of these trees in San Marino is along St. Albans Road where, each December, residents decorate the deodars with lights.

The deodar has dark gray green needles clustered on short spurs along the branchlets. By late summer large, ovoid, silvery cones can be seen growing upright on the branches. The cones are unsuitable as dry decorations because they shatter easily.

The deodar cedar grows slowly but will become quite tall and spread to cover several hundred square feet when full grown. Mature specimens are Christmas-tree-like with uncrowded, picturesque branching and distinctive drooping tops. Once a deodar cedar is established, its cultivation requirements are minimal and its demand for water very low.

Other closely related trees that may be confused with the deodar cedar are the blue atlas cedar (*Cedrus atlantica* 'glauca') and the cedar of Lebanon (*Cedrus libani*). The commonly seen blue atlas cedar has silver gray foliage and more pendent branches than those of the deodar. The cedar of Lebanon is uncommon in our area and its needles are shorter than those of the deodar.

IN SAN MARINO: *On private properties along St. Albans Road between Monterey Road and Huntington Drive.*

LOOK FOR: *Drooping top shoots.*

Cedrus deodara

2

The Pine Family

Canary Island Pine

Pinus canariensis

The Canary Island pine flourishes on dry, exposed slopes in the Canary Islands, off the northwest coast of Africa. Young trees look gawky because the needles tend to grow in tufts spaced out along the branches. Older trees are narrowly conical or broadly columnar, with layers of upsweeping branches. They grow quickly to a height of 120 feet in Southern California.

The long, blue-green needles hang from the branches, softening the outline of this tree. Its needles are so soft that they are used in the Canary Islands to pack bananas for shipping. Its glossy brown oblong cones are up to 8 inches long.

These trees are excellent for preventing soil erosion on rocky hillsides. They tolerate drought, sandy soils and seacoast conditions, but are sensitive to temperatures lower than 20° F.

IN SAN MARINO: *Along Huntington Drive between Virginia Road and Chelsea Road.*

LOOK FOR: *Long, drooping needles.*

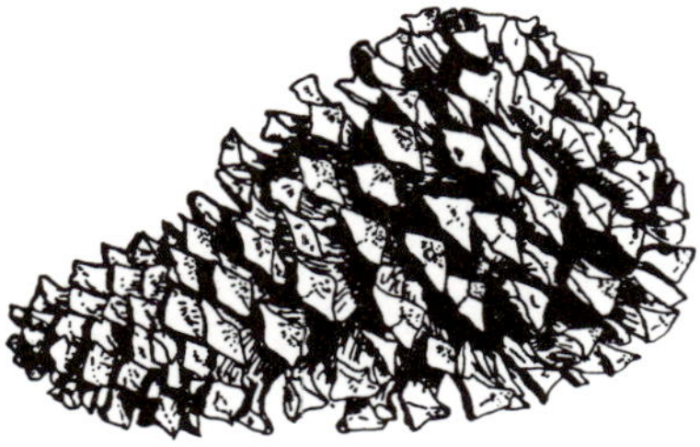

The Pine Family

Aleppo Pine

The Aleppo pine is distinguished from other large pines in San Marino by its gray green color and irregular oblong crown. Its many short, ascending branches are sparsely covered with needles and sway with the slightest breeze, making the tree seem constantly in motion.

The medium length, slender needles usually occur in pairs, but sometimes in groups of three. The Aleppo pine has brown, scaly bark and yellowish brown cones which point backwards along its branches.

Once established, this pine grows with little irrigation and is suitable for a dry, low-maintenance landscape. The Aleppo pine will grow in poor soils and windy, unsheltered areas. It is used along freeways and for erosion control. This pine also tolerates seacoast conditions. It thrives along the seashore in its native Mediterranean region.

IN SAN MARINO: *On private properties around San Marino.*

LOOK FOR: *Ascending branches on an irregular, oblong crown.*

Pinus halepensis

The Pine Family

Italian Stone Pine

Pinus pinea

A mature specimen of the Italian stone pine can be recognized by its soaring trunk covered by red-brown plates of bark and topped with a spreading flat crown. Its other common name, the umbrella pine, refers to this crown which often will have only a few large horizontal branches. In Southern California this native of the Mediterranean grows to over 80 feet.

Small Italian stone pines are deceptively different from mature specimens. Their compact, densely foliaged habit makes these trees ideal for small, living, table decorations at Christmas.

The gray green needles grow in pairs and are slightly twisted; the cones are rounded, about five inches long, and knubby. *Pinus pinea* is known for its large seed called pignoli. In the Mediterranean the seed are used in many culinary dishes. The common name, stone pine, refers to the hard seed coat.

IN SAN MARINO: *Can be seen around San Marino on the skyline.*
In Lacy Park.

LOOK FOR: *Broad, flat crown.*

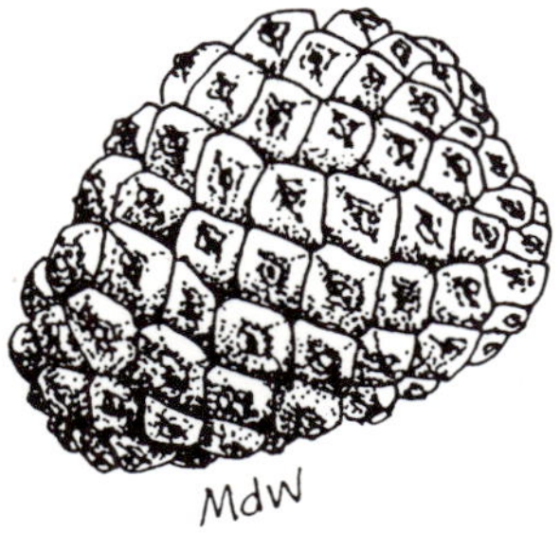

Japanese Black Pine

No other pine that grows well in Southern California is as amenable to pruning and shaping as the Japanese black pine. This native of Japan is a common Southern California landscape tree and makes an excellent bonsai subject. It can be recognized by its pairs of stiff, deep green needles and the dark brown plates on its bark.

When pruned infrequently, the Japanese black pine produces coarse growth that is awkward and unattractive compared to that of other cultivated pines. If pruned properly, it has a totally different character, displaying a handsome structure with a fine, airy foliage texture. Proper pruning involves training the top main growth for desired structure, and topping or removing the strongest new growth shoots (candles) for shaping and maintenance. In Japan, pruning is performed in the spring and fall. In Southern California the Japanese black pine has different patterns of growth, so a late fall pruning is adequate.

IN SAN MARINO: *On private properties around San Marino.*

LOOK FOR: *Stiff needles in pairs.*

Pinus thunbergiana

The Taxodium Family

Coastal Redwood

IN SAN MARINO: *A number of trees at the intersection of San Marino Avenue and Huntington Drive.*

LOOK FOR: *Tall feathery tree, tapering at the top.*

Sequoia sempervirens

Native to the moist coast ranges from the Santa Lucia Mountains south of Monterey to southwestern Oregon, the coastal redwood is California's state tree. Its distribution is not continuous, being limited to scattered groves always within the influence of coastal fogs. The coastal redwood's two closest relatives are the giant sequoia (*Sequoiadendron giganteum*) of mountainous central California and the dawn redwood (*Metasequoia glyptostroboides*) of China.

The tallest living coastal redwood is 362 feet high and about 2,000 years old. Resistance to insects, fire and rot has contributed to the longevity of members of this species. Its lumber is a pinkish red color and is of such value for its resistance to decay that large stands of *Sequioa sempervirens* along the coast of Northern California and Oregon were in danger of being totally destroyed at one time. One of the early environmental protection groups (the "Save the Redwoods" League) was organized to preserve the original coastal redwood forests. Most redwoods today are in "second growth" forests that were previously logged.

Two kinds of leaves are found on the branches of this pyramidal tree. Juvenile leaves are scalelike and appear on new shoots. The more easily visible mature leaves are lance shaped and arranged in rows along the branchlets. From a distance the foliage appears fernlike.

In the dry Southern California climate the coastal redwood does not reach the great height for which it is famous. Here it grows to 90 feet and has a tendency to suffer tip loss and bending.

Montezuma Cypress

In the San Marino area there are graceful trees that, from a distance, resemble weeping willows. On closer inspection, their pendent branches are found to be covered with ferny, needlelike foliage, indicating that these trees are conifers. Called Montezuma cypresses, they are in the same family as the redwoods.

Montezuma cypresses are native to regions of southern Texas and Mexico, where they form large stands along permanently and seasonally wet streambeds. Like the redwoods, Montezuma cypresses are long lived and can reach massive proportions. Although not exceptionally tall, they can be very broad and dense. A very large specimen of Montezuma cypress is the great tree of Tule, south of Oaxaca, Mexico. Rather than a single plant, this huge landmark may represent an entire clump of trees.

A single mature specimen of this tree would be too large for the average garden. However, the Montezuma cypress makes an outstanding landscape tree for large, open spaces such as parks or vast lawns. Its evergreen branches droop gracefully to provide cooling shelter. The foliage usually turns an attractive red in the winter prior to falling. The brief leafless period ends with an early spring production of delicate light green leaves.

IN SAN MARINO: *Two trees on the median of Old Mill Road, on the north side of Huntington Drive.*
In Lacy Park.

LOOK FOR: *Weeping form.*

Taxodium mucronatum

8

The Cypress Family

Incense Cedar

Calocedrus decurrens

This columnar member of the cypress family grows wild in California at elevations between 2,400 and 8,200 feet. It was named incense cedar because of the strong spicy aroma of its wood and crushed leaves. The wood is often used to make pencils.

The name *Calocedrus* comes from the Greek *kalos*, meaning beautiful, and *kedros*, meaning cedar. With its upward-reaching branches and rich evergreen foliage, the incense cedar is, indeed, a beautiful tree.

The casual observer may find it difficult to distinguish the incense cedar from its relatives in the cypress family. Each branchlet on this plant appears flattened, as if it had been ironed. The individual leaves resemble small scales overlapping each other on the twig. Because the leaf tips are pointed and protrude from the branchlets, they feel sharp when a hand is run lightly over the spray opposite the direction of growth.

The brown, woody cones of the incense cedar are also unusual. When closed they are almond shaped, but open cones have sections that join at the top and splay out like the pages of an open book.

Incense cedars can live to be several hundred years old. Because they grow to be fairly tall, they make good windbreaks and high screens. Once established, these trees are quite drought tolerant.

IN SAN MARINO: *Lining Rubio Drive between Melville Drive and Huntington Drive.*

LOOK FOR: *Flattened branchlets.*

Italian Cypress

Most Californians are familiar with the ubiquitous Italian cypress. Its cigar shape and dark green foliage are a common sight in local gardens.

The generic name, *Cupressus*, is thought to be from the Greek, *kuo* meaning to produce, and *parisos*, meaning equal. This may refer to the symmetrical form of the cultivated Italian cypress, although in natural populations, trees of the species are not columnar but produce spreading branches. The specific name, *sempervirens*, means always green.

In an environment of abundant water and nutrients the Italian cypress grows very rapidly, resulting in branches which bend out and down, spoiling the desired rocket shape. A small amount of water and no supplemental fertilizer will help to maintain its cigar-shaped silhouette.

The truly columnar form of the Italian cypress is commonly found growing wild in Mediterranean olive groves, hence it is often used with landscapes and architecture related to the Southern European theme. It serves as a formal border along driveways and walks. Rows of closely planted Italian cypresses are often used as windbreaks or privacy screens. Landscape architects frequently decry the overuse, misuse and loss of proportion that often result when this stately tree is used in small landscapes or is pruned to a shrub.

IN SAN MARINO: *Striking specimen on median of Sierra Madre Boulevard, one half block north of Huntington Drive. Other trees on private properties in San Marino.*

LOOK FOR: *Cigar-shaped tree.*

Cupressus sempervirens

The Podocarp Family

Fern Pine

Podocarpus gracilior

Of the many commercially available species of Podocarpus, the African fern pine is the most frequently encountered in San Marino. Though called a pine, this tree is a very different sort of conifer, more closely related to the yews. Rather than a standard dry cone, it produces a fleshy purplish seed that is often incorrectly termed a berry.

The foliage of the fern pine has a bluish or silvery aspect. Its leaves are slender and flat, similar to those of the willow. Mature trees have light brown bark and well-branched crowns. They can grow up to 70 feet in their native East Africa.

Thinning the densely foliaged limbs is worthwhile, as they may be broken off in high winds. The fern pine is often tightly pruned to form a standard topiary, as seen along Huntington Drive. Pruned, trimmed or left natural, this is one of the prettiest trees in the San Marino landscape.

IN SAN MARINO: *Lining south side of Huntington Drive between Kenilworth Avenue and Del Mar Avenue.*

LOOK FOR: *Very narrow leaves.*

The Magnolia Family

Southern Magnolia

The southern or bull bay magnolia is a native of southeastern North America and is cultivated around the world. Though it can grow to 100 feet in its native habitat, in Southern California it is generally of smaller size. Some cultivars have been selected for their compact size and are commonly used as street trees. The compact cultivar 'San Marino' originated in Lacy Park.

Magnolia grandiflora is recognized by its large, leathery, glossy leaves which have rust-colored hairs on the undersides. The tree is evergreen, but individual leaves drop throughout the year. They decompose very slowly after they have fallen.

The huge, white, fragrant flowers are produced in our climate from summer through fall. The rusty brown, fuzzy fruits which follow these flowers are oval, cone-like structures that grow erect on the branches. When they dry, they expose numerous beanlike, shiny, scarlet seed which attract birds.

An added benefit of cultivating this magnolia comes from its usefulness as cut material for decorating large areas. Boughs of *Magnolia grandiflora* are often used as Christmas greenery because of their rich green color and durability.

IN SAN MARINO: *Lining Westhaven, Bedford, and Waverly roads between Huntington Drive and Euston Road.*

LOOK FOR: *Large, glossy, oval leaves.*
Large, white flowers in summer.

Magnolia grandiflora

The Magnolia Family

Saucer Magnolia

Magnolia ×soulangeana

This small, deciduous tree is a hybrid that was developed by the French botanist, Chevalier Soulange-Bodin. He created the saucer magnolia by crossing the yulan magnolia (*Magnolia denudata*) and the lily magnolia (*Magnolia liliiflora*).

The saucer magnolia, or Chinese magnolia, is a deciduous plant that is often grown as a multi-stemmed, spreading shrub, rather than a tree. Its waxy tulip-shaped flowers appear in the early spring on bare branches. They are delicate, so they bruise easily. The flowers are white on the inside and varying shades of pink and purple (depending on the cultivar) on the outside. The tulip shape gradually unfolds to a saucer-like form which reveals the prominent stamens.

This tree is attractive even after flowering. Its oval bright green leaves are quite large and have downy undersides.

The saucer magnolia grows best when placed in full sun. Although its flowers may be damaged by frost, the plant can survive in areas that have cold winters.

IN SAN MARINO: *On private properties around San Marino.*

LOOK FOR: *Large, apple green leaves that feel velvety.*
Bare branches in winter.
Pinkish flowers in early spring.

The Laurel Family

Camphor Tree

A beautiful shade tree with a well-formed, dense crown, the camphor tree can be identified by its narrowly fissured gray bark and small, oval, glossy leaves. It is a member of the laurel family and a close relative of the source of commercial cinnamon, *Cinnamomum zeylanicum*. The foliage and bark of the camphor tree have a definite aroma of camphor. The wood makes a pleasant fragrant fire.

Although this is an evergreen tree, there is a flush of new pink- or copper-colored foliage in the spring. As the leaves mature, they turn apple green. Old leaves drop *en masse* soon after the new leaves appear. Trees are especially attractive after a rain, for water darkens the bark.

Because of its wide, round crown, streets lined with this tree are sometimes overcanopied. The camphor tree is a prolific parent, and seedlings are often found near mature trees.

IN SAN MARINO: *Lining Old Mill Road between Mill Lane and Monterey Road.*

LOOK FOR: *Apple green foliage with strong basal nerves.*

Cinnamomum camphora

The Laurel Family

Avocado

Persea americana

The American avocado or alligator pear is a member of the laurel family. Its distinctive, pear-shaped fruit has become a popular addition to the diet of North Americans. This tree is evergreen and can grow to a height of 60 feet. It is native to tropical America, including Mexico, Guatemala and the West Indies.

The wide-spreading branches and irregular crown of the American avocado are densely covered with large, dark green leaves. The tree provides good shade in addition to its crop of fruit.

Small greenish flowers appear in the early spring and are followed by fruit which mature within the year. The flesh of an avocado is oily. Its unusual flavor is very pleasant for some people and disliked by others.

The most common cultivar is the green-skinned 'Fuerte' avocado. Another cultivar, the 'Hass' avocado, has black, warty skin and originated by chance in the garden of Rudolf Hass in La Habra Heights, California, in the 1920s. All 'Hass' avocados originate from this one tree, which is still alive.

Because avocado trees have shallow roots, they must be watered frequently and are subject to uprooting by heavy winds. Roots of established trees do not respond well to disturbances such as construction for pipelines or roadways. Avocados grow best in sunny, sheltered locations that have good drainage.

IN SAN MARINO: *On the north edge of the San Marino Unified School District Office property on West Drive.*

LOOK FOR: *Large drooping leaves.*

Western Sycamore

The native western sycamore is a medium-sized, broadly columnar, deciduous tree. Its smooth light bark falls away in large thin plates. This results in a mosaic of light colors on the trunk and provides a beautiful winter sculpture when the branches are bare of leaves. The fuzzy, palmately lobed leaves and pendent, multiple clusters of ball-shaped flowering structures further distinguish the western sycamore from other trees.

As a native tree, the western sycamore requires less watering and special treatment than exotics. Homeowners may have noted a premature mid-summer browning and curling of the leaves which is caused by the fungal pathogen *Anthracnose.* This disease is most noticeable following a cool, damp spring. Considerable searching has failed so far to yield an *Anthracnose*-resistant strain.

Other sycamores of note include the eastern U.S. sycamore (*Platanus occidentalis)*, the Oriental plane tree (*Platanus orientalis)* and the London plane tree (*Platanus acerifolius)* that is commonly grown throughout England.

IN SAN MARINO: *Lining Bonita and Berkeley avenues, north of California Boulevard.*

LOOK FOR: *Flaky bark.*
Hanging groups of fruiting bodies (see illustration).

Platanus racemosa

16

The Witch Hazel Family

Sweet Gum

Liquidambar styraciflua

The American sweet gum or red gum is a native of the eastern United States, Mexico and Central America. It is a tall, broadly columnar tree and has become popular as a source of fall color.

The sweet gum can be identified by its star-shaped lobed leaves, reminiscent of the lobed leaf of maples. Sweet gum leaves alternate on the branch, unlike maple leaves which grow in pairs.

The tiny flowers of the sweet gum are produced in pendent, ball-like clusters which mature into prickly fruit called sweet gum balls. These fruit remain on the tree after the leaves have dropped, then fall to the ground during the winter and spring. They are sometimes used in dry floral arrangements.

When giving thought to a planting site for this tree, the availability of water should be considered, since sweet gums require plenty of moisture.

Though the handsome structure and foliage are reason enough to grow sweet gum trees, most people plant them because of the fine reds, purples, oranges and yellows of their autumn leaves. Indeed, the best time of year to select these trees at a nursery is after the foliage has turned color.

IN SAN MARINO: *Lining Santa Anita and Winston avenues between Robles Avenue and California Boulevard.*

LOOK FOR: *Round, prickly, hanging fruit. Colorful fall foliage.*

Chinese Elm

Although the Chinese elm has pendulous branches and a drooping silhouette, it retains the characteristic wide-spreading branches of an elm tree. This medium-sized native of Asia grows quickly and may be briefly deciduous.

The trunk of the Chinese elm has a distinctive mottled appearance. The bark falls off in flakes, exposing various shades of beige, tan and cinnamon. The small glossy leaflets are leathery, with toothed edges and asymmetrical bases.

Like other elms, the Chinese elm has winged fruit. Small, oval seed are surrounded by membranous disks that allow them to be carried by the wind, rather than fall directly to the ground. These fruit appear in the autumn and are quite showy.

The Chinese elm makes a nice shade tree because it is dense and has an appealing weeping form. If used as a street tree, the branches must be trimmed to prevent them from hanging to the ground.

IN SAN MARINO: *Northeast corner of Gainsborough Drive and Huntington Drive.*
On private properties around San Marino.

LOOK FOR: *Mottled trunk.*

Ulmus parvifolia

The Mulberry Family

Indian Laurel Fig

Ficus microcarpa

The Indian laurel fig is native to southern Asia. If left unpruned, it grows into a robust tree with a broad crown and dense foliage. Residents of Southern California will be familiar with the typical lollipop shape into which this tree is frequently pruned. It responds very well to pruning and can be successfully trimmed at any time of the year.

Its thick, shiny leaves are bright green and oval, narrowing at the tips and bases. When Indian laurel figs are used to line streets, as they are in San Marino, the glossy leaves, smooth bark and severely pruned profiles harmonize nicely with the man-made structures around them.

The Indian laurel fig is also a good choice for patios and as a shade tree. It can be placed in a tub, pruned into a hedge form or espaliered. Because it is evergreen, the Indian laurel fig does not produce seasonally large quantities of leaf litter.

It is reasonably hardy during most Southern California winters, but it will not withstand prolonged frosts and is susceptible to attack from small insect pests called thrips.

Nurseries may sell Indian laurel figs under the Latin names *Ficus retusa, Ficus nitida* or *Ficus microcarpa.*

IN SAN MARINO: *North side of Huntington Drive between Del Mar Avenue and Kenilworth Avenue.*

LOOK FOR: *Smooth bark and glossy leaves.*

Coast Live Oak

The Latin name for oak, *Quercus*, is thought to originate from the Celtic *quer*, meaning fine, and *cuez*, which means tree, reflecting the beautiful, round crowns and horizontal branches of trees of this genus.

The leaves of the coast live oak or California live oak remain on the tree all year. They are roughly oval and curve under, sometimes with spines along the edges. The thin, pointed acorns have puffy caps. Native Americans used these acorns as a food staple. The trunk has furrows separating large plates of smooth, light gray bark.

As a native of California, the coast live oak is well suited to this environment. It tolerates heat and seacoast conditions. Like all mature native California oaks, the coast live oak should not be watered in the summer. Oak root fungus (*Armillaria*) will damage the trees if the roots are continuously moist.

Large open spaces are excellent planting sites for the coast live oak. The thick, crooked branches and stout trunk combine with the dense, dark green foliage to make it an attractive specimen tree.

IN SAN MARINO: *Lining Carlisle Drive.*
On parkways around San Marino.

LOOK FOR: *Leaves that curve under (see illustration).*

Quercus agrifolia

The Beech Family

Engelmann Oak

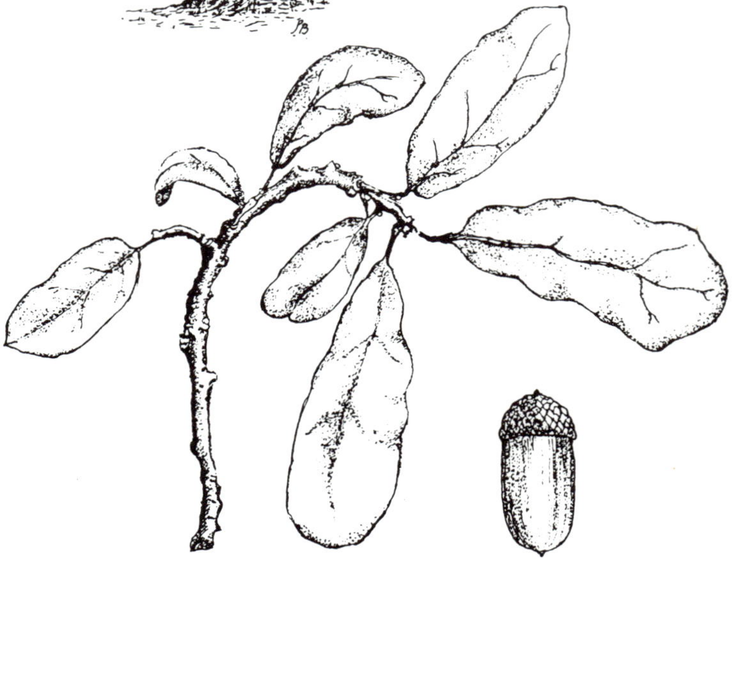

Quercus engelmannii

Variously known as the Engelmann oak, the mesa oak, and the Pasadena oak, this medium-sized evergreen tree is a native of Southern California. Like many other oaks, it has a rounded crown and horizontal, spreading branches.

The bluish green leaves of this oak are quite variable in shape. They are usually elongated and may have wavy or smooth margins. Engelmann oak acorns are oval and edible. They are half-covered with a thickly scaled cap.

These oak trees often hybridize with other oaks such as the coast live oak (*Quercus agrifolia*). For this reason, it may be difficult to identify a particular oak. In general, a true Engelmann oak will have bark which is gray and checkered with small scales. The bark of a coast live oak is usually smoother and lighter, with fewer furrows. From a distance the leaves of the Engelmann oak look slightly bluish while those of the coast live oak look deep, dull green.

Engelmann oaks are found growing naturally in the dry foothills of Southern California. They tolerate drought in city settings and can survive frosts but not prolonged cold. Because of their large spread, these trees are best planted in open spaces in parks or along wide streets.

IN SAN MARINO: *Lining Bradbury Road.*

LOOK FOR: *Bluish green foliage.*

Holly Oak

The holly or Holm oak is an evergreen native of the Mediterranean. It is a dense, medium-sized tree with stout branches. The dark gray trunk is covered with small, square scales.

The leaves resemble holly leaves, as they often have spiny edges. Unlike holly leaves, the leaves of this oak have short, yellow gray hairs on their undersides. The dark brown, top-shaped acorns are half-covered by rounded caps.

These oaks are quite hardy, surviving in areas too damp for our native oaks. While they do not have the spreading elegance of some oak trees, holly oaks are often symmetrical in shape and can be pruned to any desirable form.

Holly oaks are also used as windbreaks along the seacoast, as they can tolerate salt spray and humidity.

IN SAN MARINO: *Lining east side of Oxford Road, north of Stratford Road.*

LOOK FOR: *Dark gray bark, divided into small scales.*

Quercus ilex

The Beech Family

Valley Oak

Quercus lobata

The largest of California's native oaks is the valley oak or California white oak. It grows in fertile valleys and foothills where it dominates the landscape with stout, arching branches and a wide, rounded crown.

The large leaves have rounded lobes and pale green undersides. They have a papery texture and are shed in the fall. The reddish brown acorns of the valley oak are shaped like long thin cones and topped by warty caps. These acorns were a source of food for some Native Americans.

Valley oaks can live to be more than 400 years old. They are tolerant of drought and humid conditions, but their growth rate is slow under stress. Care must be taken when excavating near these oaks, as their roots are easily damaged when disturbed.

The valley oak is a good shade tree for parks and large gardens. Because of the chore of removing fallen leaves every autumn, this tree may not be ideal in areas where neatness is essential.

IN SAN MARINO: *North side of Lombardy Road, just east of San Marino Avenue.*

LOOK FOR: *Distinctive leaf shape (see illustration).*

Cork Oak

The cork oak is easily recognized by its extremely thick, spongy bark. In the Mediterranean, where the cork oak originates, the bark is harvested as commercial cork. The outer layer of bark is carefully stripped from the trees, which remain living and grow a new cork layer. Cork can be harvested from a single tree every decade or so for over a hundred years.

Cork oaks are evergreen, medium-sized trees that have the rounded, spreading crown typical of oaks. In old age they tend to lean and become misshapen. The bark of unharvested trees is pale gray and thick, with prominent ribs.

The shiny, lightly toothed leaves are oval and gray green above. The undersides of the leaves show prominent veins and are covered with pale gray hairs. The caps of the oval acorns have fringed scales.

The cork oak is the oak of choice for a desert setting or anywhere hot and dry. It will tolerate heat, smog and drought. When grown along streets and in parks, this tree provides a fine contrast between its light, delicate foliage and rugged bark.

IN SAN MARINO: *On Oak Knoll Avenue near Monterey Road.*

LOOK FOR: *Thick, deeply furrowed bark.*

Quercus suber

24

The Sterculia Family

Queensland Lacebark

Brachychiton discolor

The Queensland lacebark or scrub bottle tree is a native of Eastern Australia. In Southern California it is used for shade and makes a graceful avenue tree. It grows to a fairly large size and has attractive, furrowed, gray bark.

The leaves are dark green on the upper sides. Beneath, they are yellowish gray and covered with fine hairs. Most of the leaves fall just before the tree flowers. Red bell-shaped flowers appear in the late spring or early summer in clusters near the ends of the branches. The Queensland lacebark flowers most profusely after a dry spring.

The flowers are followed by golden brown, canoe-shaped pods that are covered with matted hairs. The pods can be six inches long and contain bright yellow seed set in rows, like an inside-out ear of corn. They are savored by California squirrels. The pods fall to the ground when dry. They are used by florists, for they contribute an interesting rustic look to dried flower arrangements.

This tree tolerates mild frosts and periods of drought.

IN SAN MARINO: *Lining west side of Oxford Road between Stratford Road and Euston Road.*

LOOK FOR: *Distinctive leaf shape (see illustration).*
Flowers in early summer.

Bottle Tree

The bottle tree or kurrajong is a medium-sized, symmetrical tree with dense, evergreen foliage. Its common name reflects the bottle-shaped trunk, which is stout at the base and tapers towards the top.

The Latin name, *populneus*, reflects the poplarlike quality of its leaves. They are glossy and have long stalks (petioles) that make them quiver in the breeze.

Delicate bell-shaped flowers appear in May and June. They are small and creamy white with flecks of red or purple inside. These are followed by clusters of dark brown canoe-shaped pods, containing rows of yellow seed.

The bottle tree is native to Australia, where the aborigines use its light gray, slightly furrowed bark as a source of fiber to make nets. Since it has deep roots, it is often green when other plants around it are dry and brown. Australian ranchers value the bottle tree as a source of lush, green fodder during dry spells.

The bottle tree is a good street tree for California. It provides shade, is drought tolerant, and its attractive profile enhances city streets.

IN SAN MARINO: *Lining Old Mill Road between Monterey Road and Huntington Drive.*

LOOK FOR: *Clusters of boat-shaped pods. Strongly tapering trunk.*

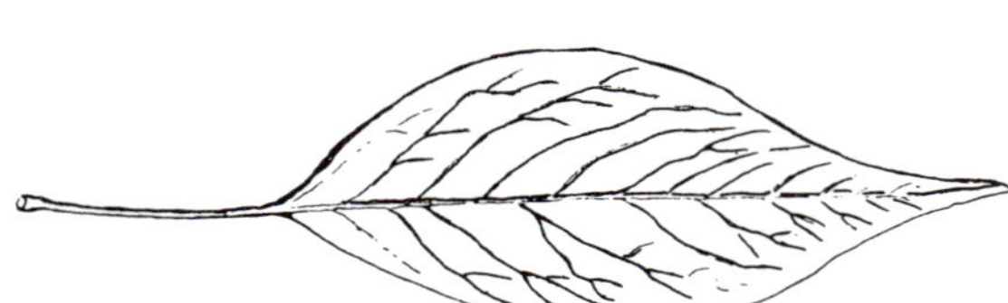

Brachychiton populneus

The Kapok Family

Silk Floss Tree

Chorisia speciosa

This South American member of the kapok family has become a popular novelty tree in the Southern California landscape. *Chorisia* is distinctive in many ways. Its slightly bottle-shaped trunk is usually green and studded with sharp, brown, barky prickles. The well-spaced branches grow horizontally from the upper trunk of this large, fast-growing tree.

As can be seen in the illustration, a single leaf comprises several leaflets arranged like the fingers on a hand. This configuration is termed palmately compound.

In late summer and autumn the tree produces large, pink flowers with white centers. A close relative, the white silk floss tree (*Chorisia insignis*), produces white flowers.

The silk floss tree is a fascinating tree to observe all year because it looks different every season. Leaves, flowers and fruit appear and are shed in an interesting progression throughout the year. The white fibers that explode from its plumb-bob-shaped fruit can be used like kapok for pillow stuffing.

IN SAN MARINO: *On private properties around San Marino.*

LOOK FOR: *Green, thorny trunk, swollen at the base.*
Flowers in early autumn.

Strawberry Tree

This evergreen member of the heath family can be a shrub or a small, multi-trunked tree. It is native to Ireland and southern Europe. The common name, strawberry tree, reflects the red, soft fruit it bears in the fall and winter. The fruit are warty in texture, giving them the appearance of strawberries. The generic name, *Arbutus*, is thought to have originated from the Celtic *arboise*, meaning rough-fruited.

Strawberry tree leaves are dark green and glossy with toothed edges. They contrast nicely with the pink or greenish white, urn-shaped flowers that appear in drooping clusters in the fall and winter. Since the fruit take a year to mature, there may be fruit and flowers on the tree at the same time.

The strawberry tree will tolerate many climates, from the desert to the seashore. Mature plants require little water in the Southern California climate. The tree should be pruned occasionally to remove basal suckers and to maintain an attractive form.

Homeowners will delight in the contrast of foliage, flowers and fruit. The red fruit fall to the ground when overripe. They attract birds and are edible to humans, although rather gritty and tasteless.

IN SAN MARINO: *Ten-foot high bush on median of Old Mill Road just north of intersection with Chelsea Road. NOTE: Intersection north of Huntington Drive.*

LOOK FOR: *Distinctive flower, fruit and leaf shape (see illustration).*

Arbutus unedo

28

The Rose Family

Carolina Laurel Cherry

Prunus caroliniana

The Carolina laurel cherry, mock orange, wild orange or cherry laurel is native to the southeastern United States. As a member of the rose family and a species of the genus *Prunus*, it is related to the true cherries, peaches, plums, almonds and apricots. Unlike many of its deciduous relatives, the Carolina laurel cherry is evergreen. It can be recognized by its dark-colored bark and dense crown of dark green, glossy leaves.

In the spring, tiny white flowers are produced on spikes. These are followed by small, hard green fruit that turn shiny black. They attract birds but are not edible to humans. These fruit eventually fall to the ground around the tree.

In Southern California, the Carolina laurel cherry is sometimes used as a tall, deep hedge. If left unpruned, it will grow into a many-branched shrub or tree. Severe pruning to clear lower branches or open the tree results in considerable suckering and will eventually produce a tree with a knurled and damaged trunk (see illustration).

The Carolina laurel cherry is highly resistant to oak root fungus.

IN SAN MARINO: *Lining Monterey Road between El Molino Avenue and Los Robles Avenue.*

LOOK FOR: *Knobby trunk.*

The Rose Family

Evergreen Pear

The common name of this small-sized tree is the evergreen pear. It is a native of Taiwan and a relative of *Pyrus communis*, the common, edible pear. The small, round fruit of the evergreen pear is hard and cannot be eaten.

The oval leaves are glossy and remain on the tree most of the year, dropping in the late fall or early winter. In some areas, they turn bright red before falling in the autumn. While the leaves are attractive, it is the flowers that provide the most beauty on this tree. In the winter or early spring, when the weather is damp and cool, the evergreen pear is covered with a dazzling display of white blossoms.

When trained into a tree form, its drooping, spreading branches and dark brown, deeply furrowed bark provide an appealing framework for the flowers. If left unpruned, it will grow as a shrub. This plant is also attractive as an espalier, although too much pruning may reduce the number of flowers.

Full sun and well-drained soil are the best conditions for growing an evergreen pear. As with many other members of the rose family, this tree is susceptible to heavy infestations of the ash whitefly.

IN SAN MARINO: *On south side of Huntington Drive between Granada Avenue and Chelsea Road.*

LOOK FOR: *Leaves with crinkled edges. White blossoms in winter.*

Pyrus kawakamii

The Pea Family

Cassia

Cassia spp.

Cassias are popular flowering trees in Southern California. Most cassias grown here are native to South America. They perform well in our strong sun and warm weather.

These shrubs or small trees are best known for their handsome clusters of bright yellow to yellow-orange flowers. The common name of shower tree refers to the falling petals. With the exception of *Cassia leptophylla*, which blooms in May and June, most cassias flower in late summer. Narrow, dark brown seed pods appear after the flowers have passed. These may be round or flat and can grow up to one foot in length.

Leaflets grow in pairs along the midveins of the compound leaves. In some plants the leaflets are narrow and cylindrical.

Several different cassias are cultivated in San Marino. The most commonly seen species are *Cassia excelsa, Cassia leptophylla* and *Cassia spectabilis.* Some cassias have recently been transferred to the botanical genus *Senna.*

IN SAN MARINO: *On private properties around San Marino.*

LOOK FOR: *Yellow flowers in early or late summer.*
Woody seed pods.

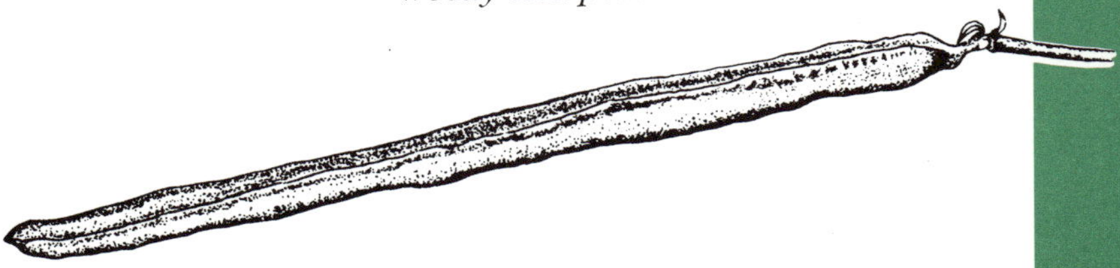

Carob Tree

This evergreen, medium-sized tree is a native of the Mediterranean. The carob is a good shade tree in parks and along streets because of its dense, round crown and uniform shape. It is also known by the common names locust bean and St. John's bread.

As is commonly the case in the pea family, the carob has compound leaves. They consist of four to ten dark green, glossy leaflets with a leathery texture and wavy edges.

Inconspicuous clusters of red or greenish flowers appear on the tree in the spring. The flat, dark brown pods, which mature early in the fall, are an attractive feature of this tree. The carob seed inside these pods are eaten like peanuts in some countries. It is thought that these hard seed were used as the original "carats" to quantify the weights of precious metals and stones. The sweet pulp inside the pods can be chewed as a kind of candy or gum. It is this pulp, in powder form, which is sold as carob, a chocolate substitute.

Once carob trees are established, they prove to be drought tolerant and resistant to light frosts.

IN SAN MARINO: *Along Ridgeway Road north of Huntington Drive.*

LOOK FOR: *Ropey, gnarled lower trunk.*

Ceratonia siliqua

The Loosestrife Family

Crape Myrtle

Lagerstroemia indica

Depending on the way it is pruned, the crape myrtle will grow as a shrub or a small tree. In Southern California, it is commonly trained as a single-trunked standard, prized mainly for its colorful early summer flowers and beautifully smooth, light-colored bark. This deciduous native of Southeast Asia is most effectively used in a formal, full sun setting, though it can also be planted in border or woodland landscapes.

Flower colors range from lavender through the pinks to burgundy and deep red. Flowers can also be white. The petals are crinkled and fringed, like crepe paper. In a good year, the flowering is so profuse that the tree looks like one huge blossom. Woody capsules grow on the branches after the blooming period.

One problem commonly encountered with this tree is mildew on maturing foliage and flowers. It is intensified in humid weather or near the coast. Mildew-resistant cultivars are the best way to deal with this problem in Southern California. Yearly winter pruning is recommended to retain a clean, open structure, though less well-maintained trees are still attractive.

IN SAN MARINO: *Lining north side of Huntington Drive between Palomar Road and Winston Avenue.*
Large, multi-trunked specimen on northeast corner of Chester Avenue and Chaucer Road.

LOOK FOR: *Light-colored bark.*
Flowers in summer.

Bottlebrush

Both the Latin name and the common name, bottlebrush, reflect the most characteristic feature of these plants. The flower petals are tiny and inconspicuous, while the stamens (the male parts of the flower) are prominent and attractively colored. The name *Callistemon* originates from the Greek *kallistos,* meaning most beautiful, and *stemon,* meaning stamen. The flowers can be brilliant red, lavender, pale yellow or apple green. Flower color and time of flowering depend on the species and cultivar. Some bottlebrush plants flower all year; most flower in the spring and fall.

The lance-shaped leaves of some Callistemon plants are an attractive copper or pink color when young. As they mature, they turn green or gray green.

Because bottlebrushes are available in a wide variety of species and cultivars, their form may vary from bushy shrubs to small trees. Some available species are the shrubby *Callistemon citrinus* and the weeping *Callistemon viminalis.*

These hardy plants are drought tolerant and survive well in smog, heat and wind. As a specimen plant or an informal screen, the bottlebrush is an excellent choice.

IN SAN MARINO: *Lining north side of Euston Road between Virginia Road and Hampton Road.*

LOOK FOR: *Flowers resembling bottlebrushes in spring and fall.*
Woody capsules near branch tips in summer and winter.

Callistemon spp.

The Myrtle Family

Lemon-Scented Gum

IN SAN MARINO: *Northwest corner of San Marino Avenue and Huntington Drive.*

LOOK FOR: *Smooth, light-colored trunk.*

Eucalyptus citriodora

Perhaps one of the most beautiful and graceful trees in Southern California is the fast-growing lemon-scented gum. The smooth, light-colored bark and open erect habit give this tall native of Queensland, Australia, an airy, elegant appearance.

From a distance, the foliage seems to grow in large clusters, like clouds, attached together with upward-sweeping branches. Upon closer inspection, the individual leaves appear sickle shaped and, when crushed, give off a pleasant lemon aroma. The bark is also fragrant. Oil from the leaves is used to perfume soap.

The most handsome and easily recognizable feature of the lemon-scented gum is its bark, which on mature trees is smooth and pale. Its color ranges from ivory to pinkish gray, depending on the time of year.

Inconspicuous cream-colored flowers appear in the early spring and fall and are followed by small, brown, urn-shaped capsules.

The lemon-scented gum grows well in the Southern California climate. Immature specimens will not withstand temperatures below 20° F., so gardeners in cooler areas may have trouble growing this tree. Because it is easily damaged by winds, the top growth should be thinned periodically.

Another eucalyptus, the blue gum (*Eucalyptus globulus*), is very common in Southern California. It is a tall, dense tree that is used as a windbreak. The bluebush (*Eucalyptus macrocarpa*) grows as a small tree or sprawling shrub. It has silver blue foliage and large pink or orange flowers. The manna gum (*Eucalyptus viminalis*) is a tall tree with graceful, drooping branches.

Cajeput Tree

This evergreen tree is small to medium sized in Southern California but grows quite tall in its native Australia. It is known by the common names cajeput tree, broad-leaved paperbark, bottlebrush, broad-leaved tea tree, punk tree and swamp tea tree.

The spongy, white bark of this tree peels off in large sheets. The name "paperbark" reflects this characteristic. The generic name *Melaleuca* originates from the Greek *melos*, meaning black, and *leukos*, meaning white. This could refer to the black specks covering the white bark.

The lance-shaped, leathery leaves are aromatic when crushed and become tinged with purple if exposed to frost. The cajeput tree has a bottlebrushlike arrangement of flowers that form at the tips of the branches in late summer. The flowers are fragrant and usually white, but can be pink or purple. The fruit are persistent, almost square, woody capsules that grow close to the twigs.

The cajeput tree grows in Australia in moist, boggy areas and has become naturalized in the swamps of Florida. Dry Southern California conditions and clay soil are also tolerated by this vigorous tree. It survives the smog and wind conditions that exist in this area.

IN SAN MARINO: *Several specimens on southeast corner of Chelsea Road and Huntington Drive.*

LOOK FOR: *White, shredding bark.*
White bottlebrush flowers in late summer.

Melaleuca quinquenervia

36

The Protea Family

Silk Oak

Grevillea robusta

Commonly known as the silk oak, this evergreen, medium- to large-sized tree is not a true oak. It was introduced into California in the late 1800s from eastern Australia. The "silk" part of its name may have arisen from the shiny, silklike quality of its wood or from the silken hairs on the underside of its foliage.

The leaves of the silk oak, with their feathery texture and lobed outlines, resemble the fronds of coarse ferns. Each leaflet curls under along the edges, adding to the soft texture. The leaves are so attractive that this tree is often grown as an indoor foliage plant. It survives well for several years indoors but grows quickly and will eventually outgrow any container.

In spring and early summer, mature silk oaks are covered with one-sided, brushlike clusters of golden orange flowers. The flowers occur at the tips of nearly leafless branchlets. They are filled with nectar so they attract birds and bees.

The silk oak is a good shade tree for Southern California because it can tolerate dry, interior climates as well as moist, coastal conditions. It makes a handsome background or ornamental tree for a large garden or park.

IN SAN MARINO: *Two trees on northwest corner of California Boulevard and Santa Anita Avenue.*

LOOK FOR: *Distinctive leaf shape (see illustration).*
Orange flowers in early summer.

The Soapberry Family

Carrotwood

This evergreen tree is known as carrotwood in Southern California and as the tuckeroo or akee tree in its native Australia. The name carrotwood reflects the orange color of its wood.

Inconspicuous white or off-white flowers bloom in the fall in fragrant clusters. In early summer, the showy orange seed capsules are relished by birds. The sticky capsules fall to the ground when ripe.

The deep green, glossy leaflets, smooth bark, and broad, rounded form make this slow-growing, medium-sized tree a popular choice. Indeed, it is a common tree in San Marino.

Periodic thinning of the carrotwood may be necessary because of its susceptibility to wind damage. This tree has deep roots that enable it to withstand drought conditions. Trees of a mature size can be successfully planted from nursery grown stock or transplanted. Temperatures lower than 22° F. may damage the carrotwood.

IN SAN MARINO: *Interspersed with queen palms along Palmas Drive.*
Along many San Marino streets.

LOOK FOR: *Glossy, slightly curled leaflets.*
Smooth trunk.

Cupaniopsis anacardioides

The Soapberry Family

Golden Rain Tree

Koelreuteria spp.

Golden rain tree and Chinese flame tree are appropriate common names for species of the genus *Koelreuteria.* They describe the shower of yellow petals that fall from the tree in the summer and the papery pods that follow the flowers. *Koelreuteria* trees are small to medium sized, with compound leaves that turn yellow and drop in the autumn.

The leaflets are egg shaped, with pointed tips and toothed margins; however, it is the flowers and fruit that catch the eye. The bright yellow flowers are small, but are arranged in large loose clusters at the tips of the branches. Three-sided seed pods follow in late summer. At first these pods are green, but later turn red, deep rose, orange, or salmon color, depending on the species. Because of their interesting color and texture, they are used in flower arrangements.

Koelreuteria can tolerate a wide range of conditions. Since the roots of these trees are not invasive, the soil beneath them can easily be cultivated. These natives of eastern Asia make wonderful specimen plants, even in winter, when their twisting, open branches set a striking scene in gardens or along avenues.

Koelreuteria bipinnata, Koelreuteria elegans and *Koelreuteria paniculata* are the three commonly grown species.

IN SAN MARINO: *South side of Virginia Road, just west of Shenandoah Road (on edge of Lacy Park).*

LOOK FOR: *Dry papery pods.*

The Cashew Family

California Pepper Tree

This medium-sized evergreen tree is a member of the cashew family. Although it is native to Peru and Chile, it has been grown for so long in the California area that it has acquired the common name of California pepper tree. It is also sometimes called the Peruvian pepper tree. It is recognized by its drooping silhouette and gnarled trunk. The trunk acquires its contorted appearance as it matures. Its rough, scaly bark is brownish gray in color.

Shiny, bright green leaves grow from hanging branchlets. These leaves are composed of many small, lance-shaped and sometimes toothed leaflets. A spicy aroma emanates from the leaves when they are crushed.

In the early summer, clusters of tiny, creamy white flowers hang from the ends of the branches. These flowers give way to bunches of pink or rose pea-sized berries which hang from female trees. The berries are quite showy and may remain on the trees throughout the winter.

Dry California landscapes are relieved by the fresh green foliage and soft texture of the California pepper tree. This fast-growing tree can survive in difficult soils and dry conditions. The far-reaching roots, which allow it to survive in drought, become a problem in a small garden as they hinder the growth of other plants.

IN SAN MARINO: *North side of Old Mill Road near Mill Lane.*

LOOK FOR: *Drooping, bright apple green leaves.*

Schinus molle

The Citrus Family

Sweet Orange

Citrus sinensis

Few people will fail to recognize the cultivated sweet orange, a small evergreen tree with aromatic leaves and fragrant white flowers. Native to Southeast Asia, oranges became popular in Europe in the 17th century. Large, heated, glass-walled buildings called "orangeries" were used to house orange trees during cool European winters. The trees were in huge containers that could be wheeled outside for the summers.

'Valencia', the juice orange, is the most commonly grown cultivar in California today. Fruit developing from the spring flowers of this cultivar turn orange by winter and are sweet by May.

Oranges should be harvested with care as the branchlets may be covered with short spines. Many cultivars of orange are available for homeowners, including dwarf forms for smaller gardens.

Oranges were first cultivated in Southern California in 1769 when they were introduced to San Diego by Spanish missionaries. At the beginning of this century acreage that would become the city of San Marino was almost fully planted with orange orchards.

The maintenance required to keep orange trees healthy and productive varies with the cultivar chosen. Rich, well-draining soil, moderate amounts of water, and a sunny location are required by all orange trees.

IN SAN MARINO: *On private properties around San Marino.*

LOOK FOR: *Round, orange fruit, white fragrant flowers. Dense, glossy, dark green foliage.*

The Olive Family

Shamel Ash

The Shamel ash, evergreen ash or Mexican ash is one of the largest cultivated ashes. Because this Mexico native has resistance to oak root fungus, it was heralded earlier in this century as the shade tree of Southern California's future. While it does provide abundant shade after only ten years, its aggressive growth and eventual large size can lead to problems in confined spaces, although smaller, slower-growing cultivars are now available for homeowners with limited space. Ash whitefly infestations have caused many people to consider other trees for shade near patios and walkways. The copious quantity of honeydew produced by these insects covers surfaces making them sticky and unattractive.

Most ash trees have similar characteristics. The leaves are paired on the stems and each leaf is composed of five to nine leaflets (pinnately compound, see illustration). The buds are cinnamon colored and have a velvety texture. In late summer the fruit mature in large, drooping clusters. They are winged, like maple seed, but each fruit is single (rather than paired as in maple).

Ashes are grouped by botanists in the same family as olive, jasmines and tea olive (*Osmanthus*).

IN SAN MARINO: *Lining Charlton Road.*
On private properties around San Marino.

LOOK FOR: *Large tree with foliage forming a rounded outline (see illustration).*

Fraxinus uhdei

The Olive Family

Olive

IN SAN MARINO: *North side of Carlaris Road between Ramiro Road and El Molino Avenue. South side of Wilson Avenue near Valentine Place.*

LOOK FOR: *Gnarled lower trunk.*

Olea europaea

The common olive is in the same family as the ashes and the lilac. Its Latin name, *Olea*, is the ancient Roman word for oil. Olive oil is pressed from the pulp of the fruit, not from the seed as with other oils. Most of the olives grown commercially are pressed for oil; only a small percentage are eaten whole.

A mature olive tree provides a pleasing contrast in appearance between trunk and foliage. The gray trunk is contorted and gnarled, while the soft, silvery green leaves look airy and light on the upward-growing branches. Fragrant, inconspicuous, off-white flowers appear in the early summer. These are followed by fruit which take a year to ripen. Olives are not edible when fresh. They must be cured or pickled to leach out their bitter taste.

Southern California has an ideal climate for olive trees, as they can tolerate extreme heat and drought conditions. Our dry air and hot sun simulate the conditions of their native Mediterranean region. Olives are medium-sized, fairly slow-growing trees. They can live to be over a thousand years old and can be successfully transplanted at any age.

The olive tree makes a fine ornamental tree in a garden because of its distinctive appearance and rugged beauty. It provides a point of interest as well as pleasant shade. There are fruitless cultivars or sprays to prevent fruit set for homeowners who wish to avoid coping with fallen fruit.

Jacaranda

Jacaranda is a genus of small- to medium-sized trees from tropical America. Most familiar to us is *Jacaranda mimosifolia*, native to Brazil and Argentina. Sometimes the tree is misidentified and sold under the name *Jacaranda acutifolia*, which is a much smaller tree with different foliage.

Jacaranda is readily identified by its feathery, twice-compound leaves that are paired along the stems. The leaves are deciduous, shattering and falling from the tree in late winter. If allowed to accummulate and decompose, the leaf litter may produce an aroma that some find disagreeable.

Most conspicuous of this tree's features is the profusion of purple trumpet-shaped blossoms, first appearing in May and continuing sporadically throughout the summer. Flowering may precede or coincide with new leaf development, depending on the amount of moisture the tree has had.

The jacaranda can have an awkward habit, branching low and yielding crossing leaders. Pruning may be necessary when the tree is young to produce a well formed specimen.

In his book, *Exceptional Trees of Los Angeles*, Donald R. Hodel describes a jacaranda tree on Los Robles Avenue in San Marino as the "most outstanding specimen in the area".

IN SAN MARINO: *Lining Banning Way.*
Lining Monterey Road between Old Mill Road and El Molino Avenue.

LOOK FOR: *Very small leaflets (see illustration).*
Purple flowers in early summer.

Jacaranda mimosifolia

The Bignonia Family

Golden Trumpet Tree

Tabebuia spp.

Native to Brazil, this small- to medium-sized tree is related to the beautiful jacaranda tree. Showy, early spring flowers are displayed on its graceful, wide-spreading branches.

Each leaf on the *Tabebuia* is composed of five leaflets attached to the leaf stalk at a single point. Leaves with this arrangement are called palmately compound. The leaflets are a shiny olive green with prominent veins.

In late winter or early spring this tree loses some or all of its leaves. Large clusters of two-inch-long yellow, pink, or lavender trumpet-shaped flowers grow on the bare branches in a spectacular display. New leaves and long, hanging seed pods follow the flowers.

Tabebuias bloom when quite small but grow fairly slowly. They grow best in full sun, with fairly moist soil, and are drought tolerant once mature. As a street, specimen or garden tree, the *Tabebuia* will delight the eye for many years.

The golden trumpet tree, *Tabebuia chrysotricha*, has yellow flowers. *Tabebuia avellanedae* and *Tabebuia impetiginosa* are close relatives of the golden trumpet tree, but they have pink- to lavender-colored flowers.

IN SAN MARINO: *Two trees on north side of Old Mill Road at Los Robles Avenue.*

LOOK FOR: *Distinctive five-part leaves (see illustration).*

Guadalupe Palm

The common name for this tree is the Guadalupe palm, reflecting its origin on the Guadalupe Islands off the coast of Mexico. It grows slowly reaching a height of about 30 feet.

Hairy filaments can be seen at the top of its brown, unevenly ringed trunk. The long, thin leaf stalks are free of spines. They support stiff, fan-shaped, light green leaves that can reach 6 feet in width. The slender yellow flowers hang in large clusters and emerge from between the bases of the leaves.

The conspicuous globular fruit are about 1 inch in diameter and dark brown to black in color. As its Latin name *edulis* suggests, the fruit of this palm is edible. It is, however, quite tasteless.

The Guadalupe palm tolerates beach and desert conditions and grows best in well-drained, periodically dry soil.

IN SAN MARINO: *South side of Bedford Road, near Durklyn Court.*

LOOK FOR: *Hairy filaments at the top of the trunk.*

Brahea edulis (Erythea edulis)

The Palm Family

Jelly Palm

Butia capitata

Native to Brazil and Argentina, this palm is known by the common names Butia palm, pindo palm and jelly palm. It is a small tree which grows slowly.

The tightly arching leaves are composed of leaflets attached to a central midrib in a featherlike pattern. The leaf bases often adhere to the plant, giving the trunk an interesting texture. The attractive blue green color of its leaves is enhanced when the jelly palm is provided with a background of deep green foliage.

In the autumn, the yellow orange fruit hang in clusters that may weigh 75 pounds. The inch-long oval fruit drop to the ground when ripe. This edible fruit can be used to make jelly, hence the common name of jelly palm.

It is an excellent palm for hot, dry, inland conditions, as it needs little water.

IN SAN MARINO: *On Allen Avenue, south of Lombardy Road.*

LOOK FOR: *Strongly curved, bluish green leaves.*

Canary Island Date Palm

Although its fruit are not edible, the Canary Island date palm is a close relative of the true date palm, *Phoenix dactylifera.* As its name suggests, it is a native of the Canary Islands.

The stout, dark brown trunk of this palm has an uneven texture due to the slow shedding of leaf bases. It may be covered with a diamond pattern from the leaf scars.

The glossy, deep green leaves form a dense, even crown. These leaves or fronds have stiff, lance-shaped leaflets extending from a central midrib in a featherlike pattern. The most basal leaflets of each leaf are quite sharp pointed and stiff. In the fall, the orange, broomlike tufts of fruit can be seen hanging below the fronds of female Canary Island date palms.

Because of its large size, this palm is not suitable for small gardens. Renowned for their uniformity of growth, Canary Island date palms look attractive when planted in rows along wide streets.

IN SAN MARINO: *Lining California Boulevard between San Marino Avenue and Sierra Madre Boulevard.*

LOOK FOR: *Stout, dark brown trunk.*

Phoenix canariensis

The Palm Family

Queen Palm

Syagrus romanzoffiana

The queen palm can be recognized by its gracefully arching, plumelike fronds and its gray, ringed trunk. Native to Brazil and Argentina, this medium-sized, fast-growing palm is a common sight in Southern California. The botanical name of the queen palm has recently been changed from *Arecastrum romanzoffianum* to *Syagrus romanzoffiana,* though either name may be used.

Its leaves or fronds are composed of many soft, gray green blades attached to a central midrib. The queen palm may have as few as five or as many as thirty such leaves, depending on its nutrition and pruning. Scars are formed on the trunk when the leaves drop. This results in bands of light and dark gray on the smooth bark.

The flowers of the queen palm are small and range in color from cream to chartreuse. They hang in a cluster on a three-foot stalk. The pale yellow, fleshy fruit ripen in June and have a sweet taste.

As an accent tree in a garden or lining an avenue, the queen palm is an attractive addition to the plant community of Southern California.

IN SAN MARINO: *Along Palmas Drive.*

LOOK FOR: *Soft, feathery leaves.*

California Fan Palm

The California fan palm, desert fan palm or petticoat palm is the only palm native to California. The Latin name *Washingtonia* honors George Washington. The California fan palm grows to 75 feet with a spread of up to 20 feet at the crown. The stout trunk is gray and smooth.

The large, fan-shaped, pleated leaves are slit into many pointed segments. They can reach a width of 7 feet. The name *filifera* refers to threads of leaf material which hang from the edges of the leaves. If the dead leaves are left on the tree, they form a shaggy, tan skirt (often called a petticoat) which may cover the trunk to the ground. In the city, these skirts are usually removed because of pest and fire hazards and the danger of falling fronds during wind storms. The edible black seed taste like sweet dates.

The desert oasis is the natural habitat for this palm, as it needs a permanent source of water. California fan palms are cultivated in parks, large gardens and along wide streets.

IN SAN MARINO: *Lining Allen Avenue, north of California Boulevard.*
Hilliard Drive has both California and Mexican fan palms.

LOOK FOR: *Stout trunk that resembles concrete.*

Washingtonia filifera

The Palm Family

Mexican Fan Palm

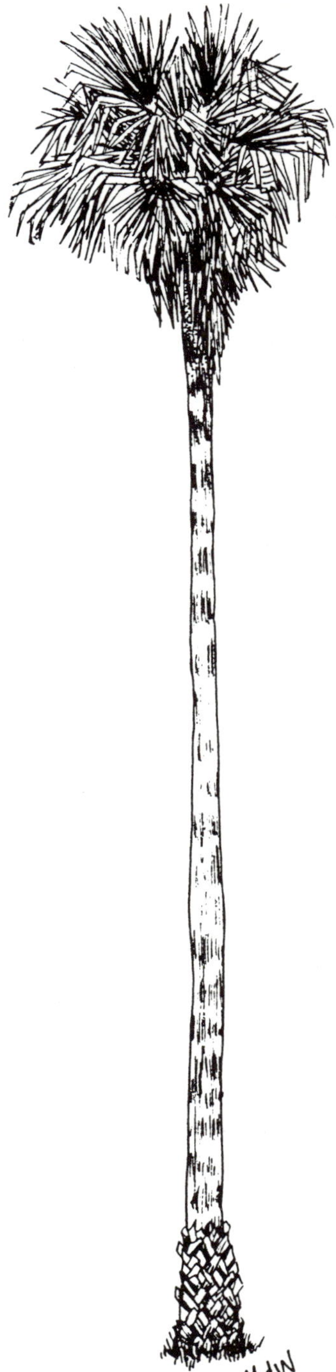

Washingtonia robusta

This native of Mexico is commonly called the thread palm or Mexican fan palm. Easily the tallest palm in our area (100 to 120 feet), this lanky tree towers above many San Marino streets, inspiring the common name, sky duster. The lollipop crown of foliage and narrow, swaying trunk distinguish the Mexican fan palm from its close relative, the California fan palm.

In proper botanical terms, the Mexican fan palm leaf type is described as palmate. At the end of the long, narrow leaf stalk, the blade fans outward with veins spreading from the point of attachment. The resulting leaf shape and lobing like the palm of a hand is the source of the descriptive term. Indeed, the common name "palm" derives from this association.

Dead leaves may remain attached to the trunk and collapse downward to form a layered, tapered petticoat. If not removed manually, this petticoat is eventually lost by the increased buffeting from winds as the tree grows and the crown reaches greater heights.

Mexican fan palm seed germinate under quite adverse conditions. Young plants are common in gutters and hedgerows.

IN SAN MARINO: *Lining Virginia Road between Euston and Lorain roads. Hilliard Drive has both California and Mexican fan palms.*

LOOK FOR: *Very tall, spindly palm.*

Suggested Reading for Further Information

Clark, D.E., editor. *Sunset New Western Garden Book.* Menlo Park, Calif.:Lane Publishing Co., 1979.

Elias, T.S. *The Complete Trees of North America: Field Guide and Natural History.* New York: Van Nostrand Reinhold Co., 1980.

Hodel, D.R. *Exceptional Trees of Los Angeles.* Los Angeles: California Arboretum Foundation Inc., 1988.

Hora, B., editor. *The Oxford Encyclopedia of Trees of the World.* New York: Oxford University Press, 1981.

Hudak, J. *Trees for Every Purpose.* San Francisco: McGraw-Hill Book Co., 1980.

Johnson, H. *The International Book of Trees.*New York: Simon and Schuster, Inc., 1973.

Kunkel, G. *Flowering Trees in Subtropical Gardens.* The Hague: Dr. W. Junk B.V., Publishers, 1978.

Lewis, M.B. *Interesting California Trees of Forest, Desert and City.* Pasadena, Pacific Book and Printing, 1983.

Macoboy, S. *Trees for Fruit and Foliage.* New York: Lansdowe Press, 1982.

Maino, E. and Howard F., *Ornamental Trees. An Illustrated Guide to their Selection and Care.* Berkeley and Los Angeles: University of California Press, 1955.

Rogers, M. *Trees of the West Identified at a Glance.* Los Angeles: The Ward Ritchie Press, 1966.

Whitman, A.H. *Familiar Trees of North America, Western Region.* New York: Alfred A. Knopf Inc., 1986.

Index

African fern pine 11
Akee tree 38
Aleppo pine 4
Alligator pear 15
American avocado 15
American sweet gum 17
Arbutus unedo 28
Arecastrum romanzoffianum 49
Ash, shamel 42
Ashes 43
Avocado 15
Beech family 20-24
Bignonia family 44, 45
Black pine, Japanese 6
Blue atlas cedar 2
Blue gum 35
Bluebush 35
Bottle tree 26
Bottlebrush 34, 36
Brachychiton discolor 25
Brachychiton populneus 26
Brahea edulis 46
Broad-leaved paperbark 36
Broad-leaved tea tree 36
Bull bay magnolia 12
Butia capitata 47
Butia palm 47
Cajeput tree 36
California fan palm 50
California incense cedar 9
California live oak 20
California pepper tree 40
California white oak 23
Callistemon spp. 34
Callistemon citrinus 34
Callistemon viminalis 34
Calocedrus decurrens 9
Camphor tree 14
Canary Island date palm 48
Canary Island pine 3
Carob tree 32
Carolina laurel cherry 29
Carrotwood 38
Cashew family 40
Cassia spp. 31
Cassia excelsa 31
Cassia leptophylla 31
Cassia spectabilis 31
Cedar,
  blue atlas 2
  California incense 9
  deodar 2
  incense 9
Cedar of Lebanon 2
Cedrus atlantica 'glauca' 2
Cedrus deodara 2
Cedrus libani 2
Ceratonia siliqua 32
Cherry laurel 29
Chinese elm 18
Chinese flame tree 39
Chinese magnolia 13
Chorisia insignis 27
Chorisia speciosa 27
Cinnamomum camphora 14
Cinnamomum zeylanicum 14
Citrus family 41
Citrus sinensis 41
Coastal redwood 7
Coast live oak 20
Cork oak 24
Crape myrtle 33
Cupaniopsis anacardioides 38
Cupressus sempervirens 10
Cypress family 9, 10
Cypress, Italian 10
Cypress, Montezuma 8
Date palm, Canary Island 48
Dawn redwood 7
Deodar cedar 2
Desert fan palm 50
Elm, Chinese 18
Elm family 18
Engelmann oak 21

Erythea edulis 46
Eucalyptus citriodora 35
Eucalyptus globulus 35
Eucalyptus macrocarpa 35
Eucalyptus viminalis 35
Evergreen ash 42
Evergreen pear 30
Fan palm, California 50
Fan palm, Mexican 51
Fern pine 11
Ficus microcarpa 19
Ficus nitida 19
Ficus microcarpa 19
Fig, Indian laurel 19
Flame tree, Chinese 39
Fraxinus uhdei 42
Giant sequoia 7
Ginkgo family 1
Ginkgo biloba 1
Golden rain tree 39
Golden trumpet tree 45
Grevillea robusta 37
Guadalupe palm 46
Gum,
- American sweet 17
- blue 35
- lemon-scented 35
- manna 35
- sweet 17

Heath family 28
Holly oak 22
Holm oak 22
Incense cedar 9
Indian laurel fig 19
Italian cypress 10
Italian stone pine 5
Jacaranda 44, 45
Jacaranda acutifolia 44
Jacaranda mimosifolia 44
Japanese black pine 6
Jasmines 42
Jelly palm 47
Kapok family 27
Koelreuteria spp. 39
Koelreuteria bipinnata 39
Koelreuteria elegans 39
Koelreuteria paniculata 39
Kurrajong 26
Lacebark, Queensland 25
Lagerstroemia indica 33
Laurel cherry, Carolina 29
Laurel family 14, 15
Laurel fig, Indian 19
Lebanon, cedar of 2
Legume family 31, 32
Lemon-scented gum 35
Lilac 43
Lily magnolia 13
Liquidambar styraciflua 17
Live oak, coast 20
Locust bean 32
London plane tree 16
Loosestrife family 33
Magnolia family 12, 13
Magnolia,
- bull bay 12
- Chinese 13
- lily 13
- saucer 13
- southern 12
- yulan 13

Magnolia denudata 13
Magnolia grandiflora 12
Magnolia liliiflora 13
Magnolia ×soulangeana 13
Maidenhair tree 1
Manna gum 35
Melaleuca quinquenervia 36
Mesa oak 21
Metasequoia glyptostroboides 7
Mexican ash 42
Mexican fan palm 51
Mock orange 29
Montezuma cypress 8
Myrtle, crape 33
Myrtle family 34-36
Mulberry family 19
Oak,
- California live 20
- California white 23

coast live 20
cork 24
Engelmann 21
holly 22
Holm 22
live 20
mesa 21
Pasadena 21
silk 37
valley 23
white 23
Olea europaea 43
Olive 42, 43
Olive family 42, 43
Orange 41
Oriental plane tree 16
Osmanthus 42
Palm family 46-51
Palm,
Butia 47
California fan 50
Canary Island date 48
desert fan 50
Guadalupe 46
jelly 47
Mexican fan 51
petticoat 50
pindo 47
queen 49
thread 51
Paperbark, broad-leaved 36
Pasadena oak 21
Pea family 31, 32
Pear,
alligator 15
common 30
evergreen 30
Pepper tree,
California 40
Peruvian 40
Persea americana 15
Peruvian pepper tree 40
Petticoat palm 50
Phoenix canariensis 48
Phoenix dactylifera 48
Pindo palm 47
Pine family 2-6
Pine,
African fern 11
Aleppo 4
Canary Island 3
fern 11
Italian stone 5
Japanese black 6
stone 5
umbrella 5
Pinus canariensis 3
Pinus halepensis 4
Pinus pinea 5
Pinus thunbergiana 6
Plane tree, London 16
Plane tree, Oriental 16
Platanus acerifolius 16
Platanus occidentalis 16
Platanus orientalis 16
Platanus racemosa 16
Podocarp family 11
Podocarpus gracilior 11
Protea family 37
Prunus caroliniana 29
Punk tree 36
Pyrus communis 30
Pyrus kawakamii 30
Queen palm 49
Queensland lacebark 25
Quercus agrifolia 20
Quercus engelmannii 21
Quercus ilex 22
Quercus lobata 23
Quercus suber 24
Rain tree, golden 39
Red gum 17
Redwood, coastal 7
Redwood, dawn 7
Rose family 29, 30
Saucer magnolia 13
Schinus molle 40
Scrub bottle tree 25
Senna 31
Sequoia, giant 7

Sequoia sempervirens 7
Sequoiadendron giganteum 7
Shamel ash 42
Shower tree 31
Silk floss tree 27
Silk oak 37
Sky duster 51
Soapberry family 38, 39
Southern magnolia 12
St. John's bread 32
Sterculia family 25,26
Stone pine 5
Strawberry tree 28
Swamp tea tree 36
Sweet gum 17
Sweet orange 41
Syagrus romanzoffiana 49
Sycamore family 16
Sycamore, western 16
Tabebuia avellanedae 45
Tabebuia chrysotricha 45
Tabebuia impetiginosa 45
Tabebuia spp. 45
Taxodium family 7, 8
Taxodium mucronatum 8
Tea olive 42
Tea tree, broad-leaved 36
Thread palm 51
Trumpet tree 45
Tuckeroo 38
Ulmus parvifolia 18
Umbrella pine 5
Valley oak 23
Washingtonia filifera 50
Washingtonia robusta 51
Western sycamore 16
White oak 23
White silk floss tree 27
Wild orange 29
Witch hazel family 17
Yulan magnolia 13